Ann is a potter. Have you heard
that word? A potter is an artist who
makes objects out of clay.

Clay comes from the earth. It
is found in soil in many parts of
America and the world.

There are several ways a potter can shape clay. Ann likes to use a potter's wheel. Ann uses her fingers to shape the clay.

Now Ann will sketch out a pattern for the bowl. She usually makes an animal pattern. Then Ann will scratch the pattern into the bowl.

Next, Ann will paint the bowl with
a glaze. She wants this bowl to be
greenish-gray.

Finally, Ann will bake the bowl and harden the glaze.

Ann is a teacher as well as an artist. Every summer, the government hires her to teach an art class. Her class starts early each morning and ends at sundown.

Ann thinks there is an artist inside everyone.

"Each of us can learn an art skill," she says. "Just let yourself have a good time. The rest will come!"